D. A. (David A.) Baldwin

The Family Pocket Homoeopathist

A Concise Manual of Homoeopathic Practice, for Families and Travelers

D. A. (David A.) Baldwin

The Family Pocket Homoeopathist
A Concise Manual of Homoeopathic Practice, for Families and Travelers

ISBN/EAN: 9783337210670

Printed in Europe, USA, Canada, Australia, Japan

Cover: Foto ©berggeist007 / pixelio.de

More available books at **www.hansebooks.com**

PREFACE.

The object of the present treatise is to present in a plain, condensed form, the Homœopathic treatment of all diseases that legitimately come within the sphere of family practice. It makes no pretensions to be a complete epitome of medical art; nor by any means to exhaust the resources of the Physician in any given disease.

But in the first place to notice only such diseases as may be easily recognized, and safely treated by any family; leaving the more acute and dangerous complaints for the care of the family Physician. And then, giving a few only of the prominent remedies that will in all ordinary cases suffice for a cure; or at least avoid the loss of valuable time until competent medical aid can be procured.

Being called for to supply a local demand, this object has been kept mainly in view.

TABLE OF REMEDIES.

Aconitum.

Apis.

Arsenicum.

Belladonna.

Bryonia.

Calcarea. Carb.

Cantharis.

Carbo veg.

Causticum.

Chamomilla.

China.

Cina.

Cocculus.

Coffea.

Colocynthis.

Drosera.

Dulcamara.

Hepar Sulphur.

Ignatia.

Ipecacuanha.

Lycopodium.

Mercurius Iodide.

Mercurius Vivus.

Nux Vomica.

Opium.

Phosphorus.

Pulsatilla.

Rhus tox.

Sepia.

Spongia.

Sulphur.

Sulphuris Acid.

Cholera Remedies.

Cuprum. Met.

Phosphoric Acid.

Secale.

Veratrum.

GENERAL DIRECTIONS.

Few will make any personal effort or sacrifice to retain or even regain health ; but relying upon medicine, expect to be cured in spite of the evil habits which have produced and foster disease. Medicines in themselves do not cure—they are given (as poisons) to excite nature to act ; and this may often be accomplished in other and better ways. One of the most important means of preserving and restoring health is

DAILY BATHING.—The importance of this is manifest when we recall the fact that there are nearly 3,000 perspiratory tubes opening upon the skin in every square inch of the body, making in all nearly 28 *miles*, if arranged as one continuous tube. Through the opening of these tubes, or as they are called, pores of the skin, over two pounds of effete, worn out matter, in the form of insensible perspiration, are daily carried off: besides that which is thrown off by visible perspiration, often amounting to as much more. If then frequent bathing and friction of the skin be not practised, the pores become obstructed; the effete matter accumulates within the system, and disease is the inevitable result.—Daily bathing for personal cleanliness is as necessary as daily food. In summer, water of the usual temperature may be used. In winter, unless for persons very robust, the water should be of a more elevated temperature, and used in a warm room to avoid chilliness. Baths thus taken are not followed by reaction, and consequently never weaken.

For debilitated persons, or those of bilious habit, baths of Alcohol and water are especially useful. Alcohol being a solvent of fatty matters, more perfectly cleanses the pores of the skin.

Salt water is also very useful for deli‹ate scrofulous subjects.

Other forms of bath for specific purposes, as sitz, and plunge baths; shower baths, packing, &c., although beneficial in many cases, are also capable of doing great injury, and ought therefore never to be used except by advice of the attending Physician.

EXERCISE.—The beneficial influence of moderate exercise in the open air is too obvious to need comment. Many of the neuralgic affections so prevalent at the present day, are due to the development and over stimulation of the nervous system at the expense of the muscular. Judicious exercise is often alone sufficient to remove the evil.

SLEEP.—Next in importance as an indispensable condition of health, is regularity in sleep. And not alone this, but sleep in the early hours of night. Many imagine that if they only secure the required number of hours sleep, it matters not *when* they get it. This is a great mistake. All day the tide of life flows with its feverish excitement, till near midnight it culminates; increased perspiration ensues, nature is relieved; and morning brings fresh vigor and strength for the day's duties. He whom midnight finds out of his bed, misses this restoring process, and awakes wearied and unrefreshed. Nor is this all; a fruitful source of disease too little appreciated, is the want of sufficient ventilation in sleeping apartments. The air exhaled in breathing is loaded with the same deadly poison, Carbonic Acid Gas, which is found in the bottom of wells and vaults, and which when breathed will destroy life instantly. Consequently the atmosphere of any sleeping room, unless thoroughly ventilated, becomes each moment more impure and unfit for respiration.— An adult will consume during eight hours of sleep, two hundred cubic feet of air ; so that Physiologists advise that each sleeping room should be of a capacity equal to twelve feet square and eight feet high ; and so ventilated as to allow a current of air to pass from without up through an open chimney or fire place, in order to secure a constant supply of pure wholesome air. When it is remembered that at least one third of life is passed in the sleeping apartment, it will be manifest that these considerations are of the utmost importance, both in health and sickness. The sick chamber more than ever requires free ventilation, though the patient should never be exposed to a draught.

CLOTHING.—As regards clothing especially of young children, it seems almost hopeless to expect any reform in this particular. In spite of the fact that more than half of the human race die before attaining the age of five years; in spite of the fact that a large proportion of deaths are the result of inflammatory affections, as Croup, Diphtheria, inflammation of the lungs and throat, induced by sudden colds, parents *will* continue to expose the naked chest and extremities of their little ones to all the changes of our fitful climate; and he who ventures to suggest that it is at great risk of life, is regarded as absurdly notional.

Underclothing, worn through the day, should always be laid aside and exchanged for clean, fresh garments at night.

DIET.—The diet of patients under Homœopathic treatment has regard first to the special nature of the disease, and then to whatever may antidote or inter-

fere with the action of the selected remedies. As respects the first, little of
general direction can be given, as what
may be entirely proper in one disease,
might be very prejudicial in another.—
All rich and highly seasoned food,
greasy substances, pastry, and food difficult of digestion, should be avoided.—
Of this class are Pork, Geese, Ducks,
Lobsters, Crabs and Clams, Sausages,
Cheese, and melted Butter. Spices of
all kinds: as Nutmegs, Allspice, Cloves,
Cinnamon, Vanilla, and Bitter Almonds.
Warm Biscuit or Fresh Bread less than
eight hours old. Some vegetables of a
pungent aromatic nature are medicinal
in their effects, and may interfere with
the action of remedies, and should therefore be dispensed with while under treatment. Of this class are Onions, Garlic,
Asparagus, Radishes, Horseradish, Celery, Parsley, &c. Coffee not always pre-

judicial in health, will surely antidote the
effect of some medicines. It is better,
therefore, in family practice, to abstain
from it altogether while taking medicine.

Black tea may be substituted in almost
all cases. Green tea should always be
avoided. It is seldom or never found
pure in this country. A commission of
Chemists appointed for the purpose,
reported through a London Journal,
that of twenty-four specimens of
Green Tea examined, *not one* was found
genuine, having been in every instance
adulterated or colored by artificial means.
Ferro-cyanide of Iron, (Prussian Blue,)
was the article most commonly used for
this purpose. Sometimes Indigo, Chinese
Clay, and Turmeric Powder, were found
in addition. Some species were largely
adulterated with a leaf resembling the tea
leaf, called by the Chinese "lie-tea."—
The genuine Imperial Green Tea is of a

dull yellowish color, not green ; and being reserved for the Chinese Emperor and Nobility, is not allowed to leave the country. Nearly all of the Black Teas, were found pure.

For obvious reasons, Camphor, Cologne, Hartshorn, and all artificial perfumes should be dispensed with while taking medicines.

ADMINISTRATION OF MEDICINES.

Homœopathic medicines are prepared for use either in the form of liquids, powders, or globules. Medicated globules are generally prescribed for Family use ; and these may be given in three ways : dry upon the tongue, five or six globules at a time ; or by dissolving each dose in a teaspoonful of pure soft water when given ; or a dozen of the pellets may be dissolved in a third of a tumbler of soft water ; a teaspoonful given to a child, or two teaspoonsful to an adult. The first method may be preferable for babes and very young children. But as in sickness the tongue is generally coated with impurities, the second method, that is dissolving each dose, is in almost all cases preferable. Wherever in this work, the dose and manner of giving is not otherwise stated, it will always be understood as prescribed in this way.

When liquids are used, five or six drops may be dissolved in one-third glass of pure soft water, and a dessert spoonful given to an adult, or a teaspoonful to a child.

The powders may be either dissolved and given in the same manner, as much at a time as would lie on a three cent piece; or taken each dose dry upon the tongue.

The repetition of dose will depend upon the nature of the disease, and

urgency of the symptoms.   In acute diseases, medicines will generally require
repetition about once in two hours.—
Chronic eases once or twice a day.

Special exceptions will be indicated
under their appropriate head.

Medicines may sometimes be given
alternately with advantage, that is, selecting two remedies, and changing from
one to the other, as often as desirable.
In general, however, it is better to give
one remedy at a time ; and never change
to another while improvement continues.

In all cases where remedies are given
frequently at first, gradually prolong the
interval, giving the medicine less often
as improvement progresses.

DISEASES OF THE SKIN.

Chafing.

This is peculiar to children, especially
of a scrofulous habit.   It most frequent-

ly occurs around the ears, in the folds of the neck, and joints of the arm and leg. The parts should be kept clean with tepid water, without rubbing; and powdered with scorched flour, or insert between the folds a soft piece of old linen, scorched and folded double.

REMEDIES.—*Chamomilla, Lycopodium* and *Mercurius.* Very obstinate cases may require *Sulphur.*

Give in the order named, one remedy at a time, a dose every three hours, and continue each one so long as improvement is manifest.

Chafing from lying in bed may be relieved by bathing the parts with *Tincture of Arnica,* or if the skin is broken, with the *Oil of Arnica.*

Chapped Hands.

Will be cured speedily by a few applications of Simple Cerate ; made of Spermaceti one part, white Wax three parts, and Olive oil six parts. As this difficulty often occurs in connection with impurity of the blood, *Hepar Sulphur* or *Sulphur*, a dose morning and evening, will remove the tendency to it.

Hives. Nettle-Rash.

REMEDIES.—*Dulcamara* if occasioned by cold. *Bryonia* and *Rhus* may be alternated if in damp weather, and attended with fever. *Nux Vom.* and *Pulsatilla* when arising from indigestion. *Calcarea* and *Sulphur* in obstinate chronic cases.

In acute cases, the remedy may be given every three hours. If chronic, give only night and morning.

Irritation of the Skin. Itching.

Caused by a fine rash scarcely perceptible upon the skin; differing altogether from the above. Is very troublesome and

474.......

long continued, unless relieved by treatment.

REMEDIES.—*Mercurius* if worse at night in the warmth of the bed. *Rhus tox* or *Apis*, if a burning itching; or *Sulphur* if changed to burning by scratching. Repeat the selected remedy every three or four hours.

Itch.

REMEDIES.—For genuine pustulous itch, take drop doses of the Tincture of *Sulphur*, and use Sulphur baths. *Sepia*, *Hepar Sulphur* and *Mercurius*, are important remedies.

Repeat every four hours.

Ringworm.

REMEDIES.—Give *Sepia*, one dose morning and evening, for a week; or as long as improvement continues. If anything further is required give *Sulphur* in same manner.

Scald Head.

This disease although characteristic of a scrofulous habit, is readily communicated. Avoid carefully the use of the same towel, brush and comb for others.— Cleanse daily with tepid water and a little Castile soap; and above all abstain from the use of any outward applications, which in many cases have driven the disease from the surface to internal organs, and caused speedy death. If the hair becomes much matted, a little pure Olive oil, applied at night, is not objectionable.

REMEDIES.—*Rhus tox* and *Lycopodium* if thick greenish crusts form upon the scalp, with a badly smelling discharge. *Hepar Sulphur* if spread over the head and neck, and with enlargement of the glands. *Calcarea* and *Sulphur* if the above should not be sufficient.

Repeat the selected remedy once in

three hours, using one medicine at a time, for three or four consecutive days, or while improvement continues.

Chilblains.

REMEDIES.—*Rhus tox* and *Sulphur*, alternately, at intervals of three or four hours.

Apply externally Arnica Oil, or Camphorated Liniment; or use every night a foot-bath of warm water, with Nitric Acid, in the proportion of one-half ounce Acid to one gallon of water.

Corns.

Soak the feet in warm water, remove the corn, and apply morning and evening Tincture of Iodine.

Warts.

Pare them and apply tincture of Iodine, or strong Acetic Acid; or else touch them daily with a little Creosote.

Felons.

Dr. Hill, of Cleveland, recommends the following excellent treatment, previous to the formation of matter :

Immerse the finger in a saturated solution of salt water, as hot as can possibly be borne ; then apply fine salt, wet in Spirits of Turpentine, and cover with a wet compress also wet with the Turpentine, and renew frequently.

REMEDIES.—*Mercurius*, a dose every three hours, with *Aconite* each hour in the interval.

If suppuration takes place, apply a bread-and-milk or Slippery Elm poultice, and take *Hepar Sulphur.*

Boils.

If much inflamed, apply the strong Tincture of Arnica, and take *Belladonna* and *Mercurius.* If matter forms, poultice, and take *Hepar Sulphur.* If very painful, add fifteen or twenty drops of Tincture of Opium to the poultice.

Carbuncle.

A painful malignant boil which sometimes becomes very dangerous from its tendency to mortification. If mild, treat it as a boil. If much swollen and inflamed, and of a livid hue, apply immediately a Yeast poultice, prepared in the following manner :   Moisten common bran with warm water till of a suitable consistence for a poultice ; add a tablespoonful of Brewer's Yeast, and set aside in a warm place till fermentation takes place.   Then place it between two folds of soft muslin, and apply.

Renew the poultice every few hours, and make each one fresh, as required, as it is of no service after fermentation ceases.

REMEDIES. — *Arsenicum* and *Carbo*

Veg., to be taken at intervals of two or three hours.

To complete the cure follow with *Hepar Sulphur* or *Silicea*.

ERUPTIVE FEVERS.

Measles.

REMEDIES.—*Aconitum* and *Pulsatilla* are the principal remedies, both for the accompanying fever, and to secure a free development of the rash.

If the eruption does not appear, or suddenly disappears, the danger is great and requires prompt treatment. In such case give at once *Bryonia*, and if attended with sickness at the stomach and deathly paleness, alternate with *Ipecac* every half hour. *Bryonia* should also be given for the hard dry cough sometimes following measles, or for any symptoms indicating inflammation of the lungs; alternate or follow with Phosphorus.

DIET.—Should be light, consisting of gruels, (barley or oatmeal,) farina, toast bread, rice, and gradually more nutritious. Water, not too cold, or milk and water may be given freely, a little at a time.

Erysipelas.

REMEDIES.—*Belladonna* in most cases, especially when the skin is bright red, or radiated, and attended with fever and headache. *Rhus tox* if little vesicles or blisters appear, and *Apis* if not relieved by Rhus, or if dropsical symptoms appear.

Should the eruption assume a dark or livid hue, or when it attacks the head and face with high fever, it may become very dangerous, and requires a physician's care.

Various applications are made use of to alleviate the intolerable itching and burning. In most cases the Yeast poultice is preferable; (for preparation see Carbuncle,) and is indispensable when the disease assumes a malignant form, turning black. Cranberry poultices will sometimes afford relief, as also thin slices of cold raw beef.

DIET.—Mainly farinaceous; no meat, and everything heating or stimulating should be avoided.

Scarlet Fever.

In general a very dangerous disease either in itself or its consequences, that should never be entrusted to family treatment. In mild cases the fever and sore throat will be relieved by *Aconite* and *Belladonna*. If the throat is very sore and ulcerated, alternate *Mercurius* with *Belladonna*. If the glands of the neck are enlarged, and mouth and throat ulcerated, *Mercurius Iod.* is preferable.—Gargle the throat also with fresh yeast, a table spoonful to four of water. The dropsical affections which sometimes follow, require *Apis, Hellebore* and *Sulphur.*

Chicken Pox.

REMEDIES.—*Aconite* for the febrile symptoms; *Mercurius,* while the pustules are maturing; *Coffea,* for restlessness and want of sleep.

Varioloid.

Is a form of Small Pox modified by vaccination.

REMEDIES. — *Aconite,* for the fever and congestion; *Mercurius,* alternated with *Aconite,* while the pustules are forming; *Sulphur,* while they are drying away; *Bryonia* and *Pulsatilla* if the eruption is suppressed, or does not come out well.

DIET.—Should n●t be heating ; give what water is required ; crust coffee, or black tea, gruels, farina, or dry toast.

Keep the patient lightly covered in a cool, well ventilated room.

Small Pox.

A much more dangerous form of the disease than Variola, and ought always to have the attention of a physician.   In the meantime the same remedies may be given as above.

This disease is highly contagious, and like many others of its class, much more so after the eruption is fully developed and maturing, than during the first few days.

Vaccination, which is a certain preventive so long as the system remains under its influence, does not protect indefinitely or for life ; precisely how long in any given case it is impossible to fore-

tell. With some, vaccination will protect for life, with others it will take on
every repetition. It is, therefore, safest
in all cases to renew it once in a few
years, or on the recurrence of an
epidemic.

FEVERS.

Simple Fever.

REMEDIES. — With chilliness, heat,
thirst and rapid pulse, *Aconite* alone is
sufficient : a dose every one or two hours.
Inflammatory fever may require *Belladonna* with the *Aconite.*

If chilliness predominates ; pain in
moving, *Bryonia.* Give water freely for
drink, and sponge off occasionally with
cold water, unless in perspiration.

Intermittent Fever. Fever and Ague.

REMEDIES.—If nausea and gastric
derangement are prominent symptoms,

give a few doses of *Ipecac*, then follow
on the well day, or during the interval
of fever, with *Arsenicum* and *China*, alternately, a dose every two hours.

During the fever give Aconite, or if
the chill and cold stage is most prominent with violent expansive headache,
brown-coated tongue, Bryonia, a dose
every hour.

During the cold stage, put warm applications to the feet, cover well, and
take a warm cup of tea, or some mild
form of stimulus until reaction sets in.

Bilious Fever.

REMEDIES.—Where there is much
chilliness with fever, yellowish-coated
tongue and bitter taste, *Bryonia*; if with
flushed face and throbbing headache, *Belladonna* or *Aconite*; if with diarrhœa and
pain in the bowels, *Chamomilla* and *Mercurius*; when with much derangement of

the stomach, yellowish-coated tongue, dizziness, dull heavy headache and constipation, *Nux* and *Pulsatilla.*

DISEASES OF SPECIAL ORGANS.

Headache.

Is frequently but the symptom of other disease, and may result from a variety of causes, which should be taken into consideration in selecting a remedy.

REMEDIES.—When produced by congestion or rush of blood to the head, *Aconite* and *Belladonna;* if from cold in the head, *Nux* and *Mercurius;* from constipation, *Nux* and *Bryonia* or *Sulphur;* from gastric derangement, *Nux* and *Pulsatilla;* for nervous headache, *Belladonna, Coffea;* and *Ignatia,* especially when caused by mental emotion. · *Sepia,* especially for females, and when upon one side of the head with sense of fullness and pressure or throbbing with nausea. And finally, for Rheumatic headache in damp weather, and with painful drawing in the back of the head and neck, *Rhus tox* and *Bryonia*.

Neuralgia of Head.

REMEDIES.—*Aconite* or *Belladonna*, if in and around the eye with profuse watering, and throbbing temples; *Arsenicum* and *Ignatia*, when mainly in the corner of the eye and at the root of the nose, coming and going at regular periods.

If the pain is generally diffused over the forehead, occurring early in the morning, attended with derangement of the stomach and constipation, *Nux vomica*.

Facial Neuralgia. Faceache.

This difficulty is frequently dependent upon a diseased condition of the teeth, which may require a dentist's care.

REMEDIES. — *Belladonna* for sharp darting pain under the eye and in the face with heat and redness. *Colocynth* when in the cheek bone, and aggravated by contact or pressure. *Mercurius* when affecting the entire side of the face from the temples to the teeth ; worse at night, and especially when in connection with defective teeth or sore gums.

The selected remedy may be given every hour. Apply externally a wet compress wrung from cold water, and covered with a dry cloth ; or, if cold applications aggravate, apply Hops, steamed with hot water.

Congestion, or Rush of Blood to Head.

REMEDIES.—*Aconite* when with heat and redness, or else paleness of the face, throbbing of the arteries of the neck and temple, sensation of fullness of the head, or with nose-bleed. If necessary, alternate with *Belladonna* for the same symptoms ; *Pulsatilla* if attended with dizziness on stooping, or when arising from indigestion ; *Nux vomica* if caused by constipation ; *Opium* from sudden fright or from costive habit.

Dizziness.

REMEDIES.—If with congestion and sense of fulness of the head, *Aconite* and *Belladonna ;* if from deranged stomach, *Pulsatilla* and *Nux vomica ;* when accompanied with nausea or vomiting, *Cocculus* and *Ipecacuanha.*

Falling off of the Hair.

When the result of fevers or debilitating sickness, *Calcarea, China* or *Sulphur* will frequently be of service. Take one remedy at a time for a week, a single dose morning and evening.

Wash the head with unrectified whiskey; or if losing the hair rapidly add

2

Tincture of Cantharides, in proportion of half an ounce to a pint of whiskey.

Inflamed Eyes.

REMEDIES.—For acute inflammation of the eyes, with redness of the eye-ball, and swelling of the lids: sensation of sand in the eyes, *Aconite* or *Belladonna* in alternation with *Pulsatilla*.

For inflammation of the lids, with discharge of matter, the eye-lids adhering together in the morning, *Pulsatilla* and *Sulphur*.

In scrofulous cases, *Calcarea*, *Hepar sulphur*, and *Sulphur* or *Mercurius* will be required.

In chronic cases, especially of elderly people, *Conium* is preferable.

Repeat the dose in acute cases every two or three hours. In chronic, a single dose morning and evening of the selected remedy.

Stye.

These may frequently be arrested when beginning to form by applying the strong *Tincture of Pulsatilla*. Give internally *Pulsatilla*, and follow if necessary with *Hepar sulphur*.

Earache.

REMEDIES.—*Belladonna* if with headache, heat and redness of the parts; alternate with *Pulsatilla*, especially if with discharge of matter from the ear; *Mercurius* if the pain extends into the face and head, and is worse at night.

Running at the Ears.

REMEDIES.—If with fever and inflammation *Pulsatilla* and *Mercurius*; or if in scrofulous subjects *Calcarea* and *Hepar sulphur*. If after Scarlet Fever or Measles, *Mercurius* or *Sulphur*.

Give of the selected remedy two or three times a day, and keep the ear

cleansed with Castile soap and tepid
water.

Humming in the Ears.

If the result of inflammation, the appropriate remedies for that condition will
relieve ; as *Belladonna, Pulsatilla* and
Mercurius.

If from cold, *Nux vomica* and *Causticum.*

If chronic, without other symptoms,
Calcarea, Causticum and *China.*

Give of the selected remedy two or
three times a day.

Deafness.

Is frequently symptomatic of other
diseases, which must first be removed by
appropriate treatment.

REMEDIES.—When it is the result of
a recent cold, *Mercurius* and *Causticum ;*
if after Scarlet Fever or Measles, *Belladonna, Hepar Sulphur* or *Pulsatilla ;* if

the result of suppressed eruptions, *Hepar sulphur*, *Calcarea* or *Sulphur*.

Mumps.

REMEDIES.—A few doses of *Mercurius* is generally all that is required. If very obstinate or painful *Mercurius Iod.* may be preferable. If attended with fever, *Belladonna*; or if the swelling is slow in passing off, *Dulcamara* or *Rhus tox.*

Nose-bleed.

REMEDIES.—To arrest the bleeding *Tinc. Hammamelis*, taken in drop doses, and also applied on a little cotton. To prevent the tendency to frequent bleeding, if it is attended with congestion of the head, red face, &c., *Aconite.* If in delicate and scrofulous subjects, *Sulphur*, or in weak, feeble persons, *China.*

Cold in the Head. Catarrh.

REMEDIES.—In all cases, in the commencement with sneezing, watery discharge from the nostril, or stoppage of the nose, alternate *Aconite* and *Nux vomica.* If the symptoms continue, or the discharge from the nose becomes acrid and irritating, *Mercurius.* If the head is much stopped alternate *Nux vomica* and *Merc.* If there is hot watery discharge from the eyes and nose, with frequent sneezing, *Arsenicum* is an almost certain specific; later, when the discharge becomes thick and yellow, *Pulsatilla* or *Dulcamara.*

If obstruction of the nose is a prominent symptom and long continued, *Causticum* will relieve. *Chronic Catarrh* requires the best care of a physician. *Aurum, Mercurius* and *Sulphur,* will in many cases be of service.

Toothache.

REMEDIES.—Give *Phosphorus* in all cases of severe jumping toothache in deeayed teeth. If this is insufficient, or when the face is swollen, follow with *Chamomilla*. If the pain is deep-rooting in the jaw, with sore or swollen gums, teeth sore and elongated, *Mercurius* is the specific; next to it *Sulphur*. In cases where sickness or other circumstances forbid the removal of teeth, speedy relief will be obtained in most cases of nervous toothache by the following application:

Tannin 20 grains; Gum Mastic 5 grains; dissolved in 2 drachms Sulphuric Ether. Apply on a little cotton.

Gum Boils.

REMEDIES.—*Aconite* and *Mercurius* alternately, while the abscess is forming; *Hepar snlphur* afterward.

Canker of the Mouth.

Is often dependent upon gastric derangement and an enfeebled constitution.

REMEDIES.—*Mercurius* is a prominent remedy, unless caused by Calomel or other Mercurials; follow, if necessary, with *Sulphur;* or if the tongue is thickly coated, bad taste in the mouth, headache, or constipation, *Nux vomica.* In nursing sore mouth, in addition to the above, *Sulphuric Acid.* For children, see "*Thrush.*"

DISEASES OF THE RESPIRATORY ORGANS.

Sore Throat.

REMEDIES. — For simple inflamed throat, soreness in swallowing, the result of cold, *Aconite* will generally suffice. If the tonsils are swollen, inflamed and painful, alternate *Aconite* and *Belladonna.*—In all cases of cankered or ulcerated sore throat, alternate *Mercurius* with *Aconite,* every one or two hours apart.

When tho diseaso is habitual, or in scrofulous cases, *Mercurius Iod.* is a most valuable remedy. Apply also to tho throat, tho wet compress, wrung from cold water, and covered with a dry napkin. If the tonsils suppurate *Hepar sulphur* will hasten the process.

For tho chronic form of sore throat, so common in this climate, involving tho palate, and tho entiro fauces, with pain and sensation of a lump in tho throat when swallowing, the remedies above named are useful. In addition *Rhus tox, Ignatia* and *Arsenicum* are frequently required; or *Nux vomica* if attended with indigestion or an acrid state of the stomach, which is in itself a sufficiant cause of the inflammation. Another cause of chronic sore throat, too often overlooked, is the irritation arising from decayed or ulcerated teeth. In acute cases the remedies may be repeated every hour.

In chronic, at intervals varying from three hours to one dose daily. Males subject to this disease should wear a full beard to protect them from the influence of sudden changes, and should avoid much covering of the throat.

Diphtheria.

This disease, though of comparatively recent appearance in this country as an epidemic, has for many years prevailed in Europe with its accustomed fatality. There are two varieties as ordinarily met with ; a simple and malignant form. The first symptoms are generally those of an ordinary cold : chilliness, flushes of heat, restlessness, pain in the bones, discharge from the nostrils, and sore throat. At this period the mucous or lining membrane of the throat assumes a peculiar livid or dark red color. Soon after a violent fever sets in ; the glands of the neck and throat become enlarged, and the peculiar diphtheritic deposite takes place, rapidly appearing upon the tonsils and throughout the fauces. This deposite is an exudation from the blood upon the dark red membrane, and consists of a thick, cheesy substance, of an ashy gray color, which is readily detached.— And this is one distinction between ordinary cankered sore throat and Diphtheria. Canker causes a depression or excavation upon the surface, whereas in the latter there first appear several minute grayish spots which rapidly run into each other, and are raised above the surface, so that they may be stripped off from it. Occasionally these spots remain separate ; and in this case it has been noticed that they dip deeper into the membrane, sometimes even perforating it like an ulcer; but this is rare. This deposite consists mainly of albu-

men derived from the blood, and shows
the serious nature of the malady; and
also that it is not a local but constitution-
al disease.

In malignant cases the fever changes
into, or assumes from the beginning a
low typhoid character; the deposite
changes to a dark brown or blackish
color, extending into the nasal passages,
causing an exceedingly offensive dark
colored watery or thick yellowish discharge; at times extending down the
windpipe giving rise to symptoms precisely similar to membranous Croup.—
However much Diphtheria resembles
Croup or malignant Scarlet Fever, it is
an entirely different disease from either,
presenting well defined marks of difference.

If the disease progresses, sloughing or
mortification of the parts may ensue,
causing an awfully fetid odor, and prov-

ing speedily fatal. There is at all times, however, a peculiar fetid odor, characteristic of Diphtheria, by which it may be recognized even before there is any appearance of it in the throat; just as in Measles or Dysentery.

The disease is more likely to assume a malignant character in feeble persons or those of scrofulous habit, but no more likely to attack those subject to ordinary inflamed throat or Croup.

Is Diphtheria contagious? Not in the same sense as Small Pox. There is no evidence that it can be conveyed by one person to another either by contact or the clothing. It may, however, be *inoculated* into the system as in Small Pox. The least particle of the diphtheritic matter finding lodgement upon a free mucous surface like the mouth and throat where it will be absorbed, may reproduce the disease in its most malignant form ; and a number of Physicians have lost their lives in this way.

Hence, the necessity for great care in this respect, as also in the use of spoons or anything pertaining to the patient.—There is no doubt also that particles floating in the air in a badly ventilated room may communicate the disease to other children, and as in so many instances it spreads through entire families, the safest way in all cases is to remove every other child from the sick room, or better still, out of the house. Experience shows that it is much less likely to affect adults than children.

The conclusions arrived at by the most eminent authorities are, that the disease is due to the presence of an especial virus which must first be introduced into the blood. That this may be accomplished by respiration or inoculation; that it *may* spread by the thorough pois-

oning of the air which is breathed; but
never by clothing; and that it requires
in all cases a special fitness of the system or predisposition in order to its development.

TREATMENT.—Two or three remedies
have so uniformly and effectually con-
trolled the simple form of the disease, in
this section at least, that little else is required. These are *Aconite, Rhus tox*
and *Iodide of Mercury.* Repeat them in
regular succession one hour apart. *Belladonna* may sometimes be preferable to
Aconite, where the tonsils particularly
are swollen, and the whole throat a dark
red; great sensitiveness to light and
noise, and the fever more of a nervous
type than inflammatory. Apply externally a wet compress wrung from a saturated solution of cold salt water, and
covered with a dry flannel; removing the
compress as often as it becomes dry.

The malignant form of the disease requires the immediate care of a skillful physician. Until such can be obtained, the above treatment may be followed as directed. Other remedies successfully used by the Homœopathic physician are *Bichromate of Potash, Bromine, Iodide of Arsenic, Cantharis* and *Nitric Acid.* For the debility following an attack of Diphtheria, *China* and *Cantharis* are the best remedies.

As preventives, *Belladonna* and *Rhus tox* will perhaps be of service.

DIET and REGIMEN.—This disease being of a very debilitating nature, rapidly exhausting the powers of life, it is indispensable to keep up the strength by a sufficient supply of nourishing food.— Milk may be taken freely; or if the patient is very weak give raw eggs beaten up with a little milk. In some cases eggnogg or wine whey are necessary. Beef

tea or a little oyster broth may be allowed, as also oranges, ice cream, or lemon-
ade if desired. Black tea, toast bread,
rice or farina as usual. Have every
article used by the patient, as knives,
plates, spoons, napkins, &c., carefully
cleansed before being again used. Keep
the room well ventilated and at an even
temperature day and night.

Hoarseness.

Is generally caused by cold and attended with other symptoms, as Cough and
Fever. In such cases *Nux vomica, Mercurius* or *Rhus tox* may be selected. If
chronic, or after an attack of Croup,
Hepar Sulphur, Phosphorus or *Causticum*
are preferable. Use one at a time at
intervals of three hours.

For the hoarse croupy cough without
fever, sometimes occurring in children
disposed to croup, give *Hepar Sulphur*
or *Phosphorus.*

Croup.

REMEDIES.—*Aconite* and *Hepar Sulphur* alternately at intervals varying from fifteen minutes to one hour apart, according to the severity of the symptoms. If not soon relieved and respiration becomes difficult, with a dry whistling sound, give *Spongia*.

Apply a wet compress to the throat, wrung from cold water and covered with a dry cloth, and renew as often as it becomes dry. If these means fail send at once for a physician.

Influenza,

Is a severe catarrh of the head, throat or lungs, due to atmospheric influences, and appearing as an epidemic.

REMEDIES.—In the commencement, for the chilliness, fever and sore throat, alternate *Aconite* and *Mercurius*. If with severe headache, stoppage of the nose.

dry hard cough, and sense of tightness across the chest, *Aconite* and *Nux Vomica*. If with running of the eyes and nose, hot and scalding, frequent sneezing, dry cough with oppression of the chest, *Arsenicum* will speedily relieve.

Cough.

Is generally the effect of a cold acting upon the respiratory organs. May be acute or chronic; dry, or with expectoration.

REMEDIES.—If the result of cold, dry with tickling in the throat, and soreness in swallowing, *Belladonna*. If with much chilliness, painful stitches in the chest when coughing or breathing, *Bryonia*. If with pain and irritation of the chest, worse in the open air, and in damp weather, alternate *Rhus tox* and *Bryonia*. When a violent racking cough deep from the chest with soreness, *Mer-*

curius. If with hoarseness, sensation of raw soreness, heat and fullness of the chest, *Phosphorus,* alternately with *Aconite.* For a hoarse croupy cough, *Hepar Sulphur.* For a dry hard cough with catarrh of the head, and constriction of the chest, *Nux Vomica.* For dry hoarse barking cough, with oppressed breathing, with heat and fullness of the chest, *Spongia.* When moist, with expectoration of a thick yellow mucous *Dulcamara;* or if loose and rattling when breathing and coughing, *Pulsatilla.* Coughs chronic, with fever and vomiting when coughing, *Drosera* and *Phosphorus.*

Cases of long standing, free expectoration of badly tasting mucous, in persons threatened with lung disease, *Calcarea* and *Sulphur.*

Whooping Cough.

REMEDIES.— For the febrile symptoms usually attending it in the commencement, *Aconite* and *Belladonna*.— For the cough when violent with hoarseness and vomiting of the food, *Drosera*. When convulsive, spasmodic, the breathing suspended for some time, *Cuprum*. In children with worms, *Cina*.

Inflammation of the Lungs.

Is a disease too grave to be entrusted to Family treatment. Until a physician can be procured give *Aconite* once an hour for two successive hours, and follow with *Bryonia* in a similar manner. Next to these in importance is *Phosphorus*.

Pleurisy.

This disease also, when attended with fever and inflammatory symptoms, requires a physician's care. The lighter forms of it may be relieved by *Aconite* and *Bryonia*, alternated, but giving each twice at intervals of an hour before changing. For the sharp stitching pains in the muscles of the chest, resembling Pleurisy, worse on moving and taking a deep breath, common to Rheumatic subjects, and occurring in damp weather, alternate *Bryonia* and *Rhus tox*, at intervals of two or three hours.

Congestion of Lungs.

REMEDIES.—For sense of weight, fullness and heat with palpitation of the heart, *Aconite* and *Belladonna*, alternately every hour.

If in females, caused by a suppression of their monthly periods, *Belladonna* and *Pulsatilla*.

Hæmorrhage of the Lungs.

Is generally a symptom of other disease requiring skillful attention.

REMEDIES.—If attended with heat, sense of weight and fullness of the chest,

Aconite; a dose every twenty or thirty minutes. If profuse, preceded by taste of blood in the mouth, accompanied with nausea and faintness, *Ipecac.* For a mucous expectoration mixed with blood, *Bryonia* and *Phosphorus.* For immediate effect, when the bleeding is profuse, give ten or fifteen drops of the *Tincture* of *Hammamelis* every fifteen or twenty minutes, till relieved. For weakness consequent upon loss of blood, give *China.*

Asthma.

REMEDIES.—*Ipecac,* when the attack comes on in the night with sense of constriction of the lungs, and rattling of mucous upon the chest. *Arsenicum,* if the former does not relieve, and there is great debility and exhaustion. *Bryonia,* for increased difficulty of breathing when speaking, or by every movement, with acute pains in the chest. *Sulphur*, in chronic cases, with profuse expectoration, sense of fullness and burning in the chest.

During the paroxysms the remedy selected may be given every half hour. In the interval once in three hours; or in chronic cases, a dose morning and evening.

DISEASES OF STOMACH AND BOWELS.

Derangement of Stomach.—Indigestion.

REMEDIES.——*Ipecac*, if with nausea and vomiting; from overloading the stomach; vomiting of mucous, and diarrhœa. *Nux Vomica*, where there is white or yellowish-coated tongue, bitter taste; acidity, flatulence, sense of fullness and tenderness in the pit of the stomach; headache and constipation; corres-

ponds particularly to a bilious tempera-
ment. *Pulsatilla*, for nausea; eructa-
tions tasting of the food; bitter taste;
tongue furred with a sticky yellowish
coat; pain in the stomach; bowels loose;
dizziness worse when stooping. Is
particularly indicated for indigestion,
caused by fat or greasy food, or in females.
Bryonia, for symptoms similar to Nux
Vomica; for acidity, water-brash, dry,
brown coated tongue; burning in the
stomach; chilliness and cold extremities;
severe headache in the temples, worse
when stooping; and constipation. —
Chamomilla, for derangement of the
stomach, commonly known as bilious-
ness; yellowish tint of the eyes and
skin; tongue yellow or brown, dry and
cracked; loathing of food; oppressive
pain in the pit of the stomach; flatulence
of stomach and bowels, and diarrhœa.

Particular indications for special symp-

toms are as follows : Heartburn, *Nux vom.*
and *Sulphur*—better still *Sulphuric Acid;*
Acidity of the Stomach, *Nux Vom. Bryonia, Chamomilla,* or *Sulphuric Acid;*
Flatulence, if attended with other symptoms of indigestion, and constipation,
Nux Vomica. If high under the ribs producing colic, without the wind escaping,
China. If of long standing and of frequent occurrence, *Sulphur.*

Dyspepsia.

When the preceding symptoms of derangement of the stomach become habitual, and of long standing, it is then generally termed Dyspepsia. The remedies
as above indicated are equally suitable
here. In addition, *Sulphur* may be given
for the same general symptoms as *Nux
vom.* Nausea, pain and fulness of the
stomach, belching of foul tasting wind,
acidity and water brash. *Calcarea* for

similar symptoms after *Sulphur*. *Carbo veg.* for sense of fulness and pressure after eating, nausea, water brash, bad breath, spasmodic pain in the stomach. These remedies may be taken at intervals of from six to twelve hours. Many obstinate cases are relieved by Bismuth, 1st trit., taken immediately after each meal.

Nausea and Vomiting.

Is generally a symptom merely of deranged stomach, for which the remedies are specified above. If caused by overloading the stomach, promote the vomiting by use of a little lukewarm water, and follow with *Ipecac ;* or where greasy food has been taken, and the nausea is attended with dizziness, *Pulsatilla.* For nausea and dizziness when moving the eyes or head, like sea sickness, or when caused by the motion of a swing or a carriage *Cocculus*. For vomiting of bile, greenish looking, and bitter, *Ipecac*, *Chamomilla* and *Nux vomica* are indicated.

Vomiting of Blood.

REMEDIES.—If caused by an accident or mechanical injury, *Arnica*. If from disease of the stomach, with great prostration and nausea, *Arsenicum*. In females with suppressed menses, *Pulsatilla*. If hæmmorrhage is profuse, give *Tincture Hammamelis*, ten or fifteen drops, every half hour, till arrested.

Sea Sickness.

REMEDIES.—*Nux vomica*, *Pulsatilla*, *Tabacum*, *Sepia* and *Cocculus*, are the remedies which have proved most serviceable. *Nux Vomica* is recommended to be given three or four doses before sailing. When sickness occurs, alternate *Nux* and *Pulsatilla*. If unavailing, follow with the other remedies, continuing any one that may afford relief.

A teaspoonful of Æther in a glass of Sherry Wine or Brandy, will probably relieve as speedily as anything.

Cramp of the Stomach.

REMEDIES.—*Nux Vomica* when appearing soon after a meal, with spasmodic, contractive pains, sense of weight and oppression of the chest, nausea, and especially if caused by Coffee or Stimulants. Repeat every half hour. *Chamomilla* for similar symptoms, worse at night, with great anguish and restlessness, and increased irritability. *Cocculus* when the above is not sufficient, and the pain is temporarily relieved by raising wind. *Sepia* in cases of habitual indigestion. *Ignatia* if caused by grief or mental emotions. *Pulsatilla* and *Cocculus* in females when attendant upon the monthly period.

Inflammation of the Stomach.

Is characterized by a constant, violent, burning pain in the stomach, with heat, throbbing, tenderness on pressure, and vomiting. Is sometimes caused suddenly by cold drinks while the system is over-heated. Is a very dangerous disease, requiring prompt medical aid. In the meantime a few doses of *Aconite* may be given at intervals of twenty or thirty minutes; followed by *Bryonia* in the same way.

Constipation.

This affection when not symptomatic of other disease, will generally yield to one of the following

REMEDIES.—*Nux Vomica* when there is ineffectual desire, congestion of the head and headache, gastric derangement, and especially in persons suffering with Piles. Should be given one or two doses a day, in the afternoon or evening. *Sul-*

phur may be given for the same class of symptoms after it, or in connection with it, in obstinate cases. Give a dose of *Sulphur* in the morning and *Nux vom.* at night. *Opium* when there is great torpidity of the bowels, without any desire for action. *Bryonia* for persons of bilious habits, with disposition to headache, chilliness, and gastric derangement, or after bilious fevers. One or two doses a day will be sufficient. Avoid in all cases the use of purgative medicines. If it is desirable to procure a speedy evacuation of the bowels, use an injection of tepid water with a little Castile soap rubbed into a light suds. When the difficulty is habitual, solicit a movement regularly at the same hour each day, and assist it by a careful kneading of the bowels.

For diet, use coarse brown bread, made from unbolted wheat; also fresh fruits and vegetables. Dispense with coffee, drink freely of water; avoid salted meats, cheese, and all highly seasoned food. Exercise freely in the open air.

Diarrhœa.

No disease yields more promptly than this to well selected Homœopathic remedies. Somewhat varied are the indications for their use. In many cases where diarrhœa is the result of overloading the stomach, or eating indigestible food, nature relieves itself by carrying off the offending substance through the bowels, or by vomiting. In such cases little else is required than rest. If the stomach remains irritable with nausea and vomiting, with watery, greenish, or slimy evacuations, *Ipecacuanha* will relieve.— *Mercurius* is suitable for a bilious diarrhœa; dark green, yellowish, slimy or bloody stools, very badly smelling; sharp cutting pain in the bowels, with nausea and faintness at the time of movement; urging and straining at stool; and worse in the after part of the day and night. *Chamomilla* when there is severe colic-like pain in the bowels, evacuations of a thick, greenish, chopped like appearance, yellowish, or slimy. Is especially useful in diarrhœa of children while teething. *Dulcamara* for symptoms similar to Chamomilla, but always when the result of taking cold, with yellowish, watery, or slimy discharges. *Sulphur* for the ordinary bilious summer diarrhœa, with griping pain in the bowels, as after the action of a cathartic; yellowish or slimy stools, with pressure upon the rectum; food undigested. *China* where the diarrhœa comes on immediately after eating, and consists of undigested food, with much flatulence, and colic pain in the bowels, and sense of great weakness.— *Arsenicum* for frequent watery or green-

3

ish discharges, with thirst, restlessness, rapid prostration of strength, sunken eyes, heat in the stomach and bowels; for painless, involuntary watery evacuations, with nausea and vomiting; for diarrhœa of teething children, and cholera infantum. *Veratrum* for violent cases with coldness and rapid loss of strength; alternate with *Arsenicum;* and finally, *Phosphoric acid* for watery, light colored or involuntary evacuations, attended with loud rumbling of the bowels, is a never failing specific. These remedies should be given at intervals, varying from one to three hours, according to the urgency of the symptoms, gradually prolonging the interval as improvement takes place. Rest is essential to a speedy cure.

Diet.—Avoid acids, meats and vegetables. The food should be mainly farinaceous:—Toast bread, rice, farina, &c.. with black tea.

Dysentery.

Is not a diarrhœa or looseness of the bowels; on the contrary it is character-ized by constipation, or a retention of the natural discharges. The evacuations are slimy or a bloody slime; sometimes pure blood, and attended with fever, violent cutting pain, and tenesmus, or urging, straining at stool. These symp-toms are caused by congestion and in-flammation of the lining membrane of the bowels; and accompanied with almost entire inaction of the liver. As soon as the evacuations again become bilious, with abatement of the fever, the disease is subsiding.

REMEDIES.—*Mercurius* is the great specific in this disease. Special indica-tions for its use are the mucous or bloody evacuations; nausea, urging and straining at stool, as if the bowels would be forced out; worse at night. When attended with fever alternate with it *Aconite;* or if the colic pains are severe, and discharges mixed with greenish or bilious matter, *Colocynth.* *Nux vomica* may be useful where the evacuations are small and frequent, with violent cutting pains in the bowels and excessive straining.

Give *Mercurius* every second hour, and either of the above remedies when required, once in the interval. In cases where *Mercurius* does not seem to act sufficiently, give *Sulphur* for twelve or twenty-four hours, a dose every third hour, and then resume the *Mercurius.*— When there is much discharge of blood, and severe pain, preventing rest, immediate relief may be obtained by using a starch injection with a teaspoonful of *Tincture Hammamelis* added. The whole injection should not contain more than a table spoonful, in order that it may be retained. Repeat the injection if necessary once in six hours. The application to the bowels of a wet compress, wrung from cold water, and covered with a dry cloth, is also of great service, and may be repeated as often as it becomes dry.

Diet.—All animal food and vegetables must be dispensed with, except the use of mutton broth, when there is not much fever. Stimulants are hurtful. Farina, gruels of various kinds, and black tea, or water in small quantities may be allowed. Entire rest upon the back is essential.

Cholera Morbus.

Occurs mostly in summer, and generally comes on in the night, with vomiting, purging, and pain in the stomach and bowels.

REMEDIES.—When the vomiting is prominent, commence with *Ipecac*, a dose

every twenty or thirty minutes, if necessary. If with thirst and restlessness,
profuse watery evacuations, great prostration, give *Arsenicum* in the same manner ; or if with severe cutting pain in the
bowels, increased prostration, coldness
of the extremities and cramps, alternate
Arsenicnm and *Veratrum.*

Asiatic Cholera.

This epidemic and fatal disease, requires the most prompt and skllllful treatment. As, however, it is often impossible during such an epidemic to secure at
once the services of a Physician, life may
often be saved by a knowledge of the
proper course to be pursued.

TREATMENT.—The disease is generally preceded by a diarrhœa of a day or
two standing, which if neglected, will soon
end in fully developed cholera. This is
commonly a light colored watery pain-

less diarrhœa, attended with rumbling of the bowels, and requires the use of *Phosphoric acid*, a dose every one or two hours, till checked. If a bilious diarrhœa, with griping pain in the bowels, *Sulphur* is preferable. Absolute rest also is indispensable. Should the disease progress to an attack of cholera, with vomiting and rice water discharges, *Camphor* is the first remedy in all cases. Give the strong *Camphor* spirits, (prepared one part Camphor to six parts strong Alcohol,) in drop doses, on a little sugar, or in a spoonful of iced water, every five or ten minutes; gradually prolonging the interval as the symptoms improve. If free perspiration ensues, discontinue it. Should there be no improvement within a couple of hours, then give *Veratrum*, a dose every fifteen or twenty minutes; or if there is intense thirst, burning heat in the stomach, rapid prostration of strength, give *Arsenicum* in alternation with *Veratrum ;* or if cramps are prominent symptoms, alternate *Cuprum* and *Veratrum*. For the collapsed stage of cholera, *Carbo veg.* is the principal remedy, and may be alternated with *Arsenicum.*

For the distress of the bowels, sensation of fulness, flatulence, and disposition to diarrhœa, so prevalent during such an epidemic, *Chamomilla, China,* and *Sulphur* are the best remedies. A broad band of flannel worn round the bowels is also useful.

Cuprum and *Veratrum* have been generally recommended as Preventives, to be taken a single dose on alternate days.

Diet.—For persons in health during the prevalence of cholera, change the ordinary diet as little as possible. Avoid all unripe fruit; certain vegetables, as cucumbers, squash, cabbage, green beans, melons, &c., and in general every thing which is found to disagree with the stomach at any time. Beef, mutton, good potatoes, ripe berries and peaches, constitute the best diet.

During an attack, rice or toast water, farina gruels, or beef tea, may be allowed; and for the extreme thirst, cold water in small quantities, or if this induces vomiting, bits of Ice.

Colic.

REMEDIES.—*Colocynth* for colic proceeding from flatulence which cannot be discharged; sharp violent pains, either constant or returning at short intervals, bruised feeling of the bowels, and especially if in connection with other bilious symptoms; give every fifteen or twenty minutes till relieved. *Nux vomica* for severe pain in the lower portion of the bowels, sharp cutting pains, pressing in every direction, and relieved by sitting or lying down, soreness of the abdomen and constipation. *Chamomilla* especially for children, and when attended with greenish or bilious diarrhœa. Injections of warm water, as hot as can be borne, will often give immediate relief. *China* for colic, with flatulence, worse at night.

Inflammation of the Bowels.

A characteristic of this disease is extreme tenderness upon pressure, so that even the weight of the bed clothes cannot be borne, in consequence of which, the patient lies with knees drawn up to the body; commences with chill, fever, and the usual signs of inflammation; requires at once the care of a Physician. In the meantime *Aconite* and *Belladonna* may be given alternately every hour.

Piles.

Presents itself in two forms ; as blind or flowing piles, though there is no essential difference between them ; merely a difference of degree of congestion.— The disease is almost always accompanied with or aggravated by constipation. It is also produced by the use of alcoholic stimulants, excessive use of Tea and Coffee ; the use of cathartics, particularly Aloes, Rhubarb, and Jalap ; sedentary habits, worms, or compression of the waist by tight clothing.

REMEDIES.—*Nux vomica* and *Sulphur* are most frequently indicated. One dose of the latter in the morning, and the *Nux vom.* at night; may be given more frequently in acute attacks. *Nux vomica* whenever there is constipation, ineffectual desire for movement, colic pains, pressing pain in the back, and in persons of sedentary habit. *Sulphur* for itching and burning, and the tumors moist, with sensation of weight and fulness in the rectum. *Arsenicum* when there is extreme heat and burning in the tumors and lower part of the bowels; for either blind or flowing piles. *Belladonna* for bleeding piles, with severe pain in the loins. If the bleeding is considerable, *Tincture Hammamelis* will arrest it. Inject a teaspoonful with a little starch paste, so that the whole does not exceed a tablespoonful; and take three or four drops in a little water, every twenty minutes. Applications also of Tincture Hammamelis by means of a soft cloth, will relieve the pain and soreness. Oil of Arnica, is also excellent for this purpose, and also the warm Sitz bath.

Diet.—The diet should always be such as to favor free and easy evacuation of the bowels; hence mainly fruit and vegetables; avoiding stimulants and spices of every description.

Worms.

REMEDIES.—If accompanied with fever *Aconite.* For pin worms *Ignatia* and *Sulphur ;* and inject a little pure sweet oil into the rectum where they are usually lodged. For the long round worm when seen, or where there is picking of the nose, irregular appetite, fetid breath, grinding of the teeth, colic pains, and restless peevishness, *Cina* and *Mercurius.* For the tape worm, (white, flat and jointed,) or in all obstinate cases, give perseveringly *Calcarea* and *Sulphur.*

For Diet.—Milk, meat broths and meat where there is not much fever, is better than vegetable food. Milk particularly is regarded beneficial during attacks of worm colic.

Inflammation of the Liver.

Is in most cases a severe disease, requiring the attendance of a Physician.

REMEDIES.—If fever, dry hot skin, thirst and chilliness, *Aconite.* If tenderness upon pressure, pain in the right side, worse on moving or even breathing, yellowish or brown coating of tongue, *Bryonia.* If the pain is dull, not aggravated by pressure or motion, yellowish skin, bitter taste, and yellow coated tongue, *Chamomilla.* Where the patient is jaundiced, with yellowness of the skin and eyes, bitter taste in the mouth, chilliness, aching pain, with inabilty to lie on the right side, and a clammy perspiration, *Mercurius.* When there is sharp pain in the liver, swelling of the right side, diarrhœa, with redness of the face, *China.* And finally for shooting pains, and great tenderness of the right ride, especially if accompanied with gastric symptoms, as nausea and vomiting, or sour and bitter taste, headache, high colored scanty urine, and constipation, *Nux vomica* is the remedy.

Liver Complaint.

As generally undersood, denotes a chronic inflammation of the liver.

REMEDIES. — In addition to those named for acute inflammation, *Sulphur* will be useful where Mercury in any form has been taken in excess. *Lycopodium* where the bowels are habitually constipated ; and *China* when the bowels are loose, or where the symptoms are more prominent every other day.

Jaundice.

Is caused by torpidity of the liver, and consequent deposition of the bilious coloring matter in the skin and other organs.

REMEDIES.—*Chamomilla* and *Mercurius* will generally suffice in mild cases. *Bryonia* if attended with chilliness. *Nux vomica* if with gastric derangement and constipation. *China* and *Sulphur* in obstinate cases.

Biliousness.

REMEDIES.—*Bryonia* and *Nux vomica* for chilliness, headache, weariness, bitter taste, and constipation. *Chamomilla* and *Pulsatilla* for dizziness, jaundiced appearance of the skin; tongue coated brown or yellowish, oppressive fulness of the stomach and bowels, and diarrhœa.

Use also sponge baths of alcohol and water.

Inflammation of the Kidneys.

Is attended with a dull distressing pain in the region of the kidneys; that is, on either side of the spine, between the hip and the short ribs; the pain is aggravated by stooping, coughing or by motion; and by lying on the affected side. It is attended with chill and fever, and in most cases with numbness of the thigh on the affected side; generally the secretion of urine is much diminished, and passed with much pain and burning; sometimes mixed with blood or matter. Is most liable to affect adults; and may be the result of sudden cold, of gravel, violent lifting, suppressed hæmmorrhages, or excessive use of stimulants.

REMEDIES.—*Aconite* for the fever, in repeated doses, every hour; or *Belladonna* if the pain recurs periodically; pain, stinging burning, extending to the bladder; urine scanty and very high colored, and with colic pains in the bowels. *Nux vomica* for dull heavy pain, mostly in the back; and when the result of a cessation of the accustomed bleeding of the piles, or from excessive use of liquor. *Cantharis* for sharp cutting pains, but particularly when the emission of urine is intolerably painful, a few drops at a time, with burning, stinging pain, and urine sometimes mixed with blood.

The remedies may be taken every

hour, prolonging the interval as improvement follows.

Diet—Should be light and unstimulating, mainly of gruels; and for drinks, cold water, Crust Coffee or solutions of mucilaginous substances, as gum Arabic, Slippery elm, or Flax seed.

Inflammation of the Bladder.

Is recognized by pain, heat, and tenderness over the bladder; urination difficult, painful and high colored.

REMEDIES.—Similar to "Inflammation of Kidneys." *Aconite* or *Belladonna* for the fever. *Cantharis* for the pain in urinating. *Nux vomica* and *Pulsatilla* in mild cases, or for frequent recurrence on taking cold.

Painful Urination.

Is generally accompanied with heat, frequent and urgent desire, with inability to pass but a few drops at a time. Such symptoms require *Cantharis.* If the result of a sudden cold, or the urine is bloody, alternate with it *Aconite ;* or if from suppressed Piles, or excessive use of stimulants *Nux vomica.* If the pain is in the extremity of the passage, accompanied by a mucous discharge, *Mercurius* is the remedy, followed if necessary, by *Sulphur.*

Rheumatism.

A disease best appreciated when experienced ; located nowhere ; obtruding its unwelcome visage in every hole and corner of the human dwelling ; searching diligently for, and remorselessly seizing upon every weak and unguarded point of human frailty. Sometimes enters your dwelling without a moment's warning of preparation for your distinguished visitor; wanders from room to room, and while yet congratulating oneself that it is but a kitchen visitor, is found in full possession of the parlor; consults no ones convenience; spares neither old nor young; and has no manner of respect for even the medical profession. For practical purposes the disease may be divided into two forms, acute and chronic. The acute form is generally developed after a sudden cold or check of perspiration, with chill and fever.

REMEDIES.—*Aconite* where the fever is high, of an inflammatory character; dry, hot skin, thirst and redness of face, with sharp shooting pains. If the affected parts are red and shining, with swelling, alternate with *Belladonna*. *Mercurius* when there is profuse perspiration which affords no relief; pains worse at night, and when warmly covered. *Pulsatilla* if the pains suddenly change their location, wandering from one part to another, causing swelling and redness, worse at evening, and particularly when located in the foot and lower extremities.— *Bryonia* particularly when located in the joints, sharp darting pain on the least motion or pressure, attended with stiffness and swelling, headache, thirst and sour sweats. *Rhus tox* where the pains are worse during rest, relieved by motion; brought on by cold, and worse in damp, wet weather. Alternates well with *Bryonia.*

Alcoholic vapor baths will afford relief for the acute pain, as will also the application of hot alcohol, and friction with flannel. Wet compresses from cold water, will often allay the pain and reduce inflammation, as also other forms of bath, which should only be used by advice of a Physician.

Chronic Rheumatism.

Any of the above remedies may be equally useful in the chronic form of the disease. In addition, *Sulphur* may be used in most chronic cases, where there is recurrence on every trifling exposure, when the pains become seated, affecting the joints and the limbs, relieved by external warmth, and aggravated by cold, especially if attended with gastric disturbance, loss of appetite, acidity, &c. Many obstinate cases have been permanently benefitted by a persevering use of Sulphur-water or vapor baths. *Nux vomica* when the pains are principally in the back, with stiffness and inability to move, with headache, and costive habit. Rheumatic subjects ought always, summer as well as winter, to wear flannel next the persons; to guard the feet well from dampness, and carefully avoid the sudden checking of perspiration.

Lumbago.

Is of a rheumatic nature, located in the muscles of the back and loins; comes on suddenly; sometimes with fever.

REMEDIES.—When with fever commence with *Aconite;* alternate with it *Nux Vomica* if the pains are aggravated by motion, and extend up and down the back. *Bryonia,* with chilliness, and worse in the morning. *Sulphur,* when obstinate, or of frequent occurrence. If located in the neck, and back part of the head, producing what is called "crick in the neck," *Bryonia* and *Rhus tox.* In cases produced by over straining of the muscles, use *Arnica* internally, and apply outwardly the *Oil* or *Tincture* of *Arnica.*

Sciatica.

Is characterized by a violent rheumatic pain, commencing in the region of the hip, and following the course of the sciatic nerve to the knee or foot; a neuralgic rheumatism; a very painful and obstinate disease, requiring skillful attention. *Arsenicum* and *Nux Vomica* are important remedies; or *Chamomilla*, when attended with great nervous restlessness.

Burns and Scalds.

Probably the best application for burns, no matter how extensive, is soap; envelope the parts in soft soap; or scrape Castile soap and make a paste with tepid water and apply; or make a thick lather and apply with a brush. This should be continued until the parts are entirely healed. Another excellent application is Sweet Oil or Petroleum, applied on raw cotton; or if the hand or arm, immerse it at once in milk, continuing it until the pain wholly ceases, and then bind on cotton batting. It is important to keep it from the air. Light cases of burns where the skin is not broken, may be treated with *Arnica Tincture.* Should it be attended with fever give *Aconite* and *Belladonna.*

Wounds and Bruises.

For bruises the best application is *Arnica Tincture.* Where there is reason to apprehend discoloration, black and blue appearance, do not apply cold water. This congeals the blood and produces it more certainly. Use warm water, by which the blood is rendered more fluid, and circulating more freely through the small vessels, discoloration is avoided. In wounds accompanied with bleeding, notice whether the blood is dark, issuing in a continuous stream ; or if bright red, and ejected by regular pulsations. In the former case it is from a vein, and requires nothing but pressure to stop it. Bind tightly over the wound a small compress of muslin, folded several thicknesses. If the blood issues in pulsations, it is arterial, and dangerous if not soon stopped. In such case, tie tightly with a cord, or pocket handkerchief *between the wound and the heart*, wherever that may be, until a Physician can be procured to arrest it permanently.

Sprains and Strains.

Probably nothing better can be done in all ordinary cases than to keep the parts well bathed with *Tincture Arnica;* or if the joints and tendons principally are injured, *Rhus tox* may be applied, and taken inwardly.

Stings of Insects.

For bee or wasp stings, Quinby, the celebrated Apiarist, recommends the common garden Onion, applied where the sting entered. Cut the fresh onion, and apply it to the spot, changing it every ten or fifteen minutes, till the pain and swelling disappear. Spirits of Ammonia, the common Hartshorn, is another excellent application for the sting of any Insect. For the poison of Serpents, Brandy or Whiskey taken to the extent of intoxication, is said to be the most reliable means of cure.

Poisons.

In all cases of poisoning, as soon as possible thereafter, empty the stomach by an emetic. This may readily be done by a tablespoonful of common Mustard, in half a tumbler of warm water. To neutralize poison if known to be Arsenic, Corrosive Sublimate or Sugar of Lead, use freely the whites of eggs.

If the poison is unknown, procure of any Druggist the following prescription, which will neutralize most mineral poi-

sons.   Take Calcined Magnesia, Pulver-
ized Charcoal, and Sesquioxide of Iron ;
mix in equal parts, dissolve in water and
take freely.   In the meantime use whites
of eggs.   Strong, black Coffee, will neutralize the bad effects of Opium, Nux
Vomica and Stramonium.   For poisoning by Acids, give a strong solution of
common white Soap, dissolved in warm
water.

Diseases of Females.

It is estimated that scarce one female
in ten, between the ages of fifteen and
forty, are entirely free from some form
of Uterine disorder.

One fruitful source, often urged but
not always understood, is faulty dress-
ing.   Females are apt to suppose that if
they do not take cold, or experience any
immediate evil effects from imperfectly
protecting the neck, shoulders, arms and

feet ; or from suddenly changing from the warmest woolen clothing of the day to evening dresses of the lightest fabric, that the practice is not injurious ; yet here is laid the foundation of diseases that afterward render life burdensome.

The blood is conveyed from the heart by deep seated arteries to supply all the inner organs of the body ; a large proportion of it is returned to the heart by numerous veins immediately under the skin. If then, the surface of the body be exposed to the cold air, or the extremities not sufficiently protected, the blood is driven from the surface, producing congestion of the inner organs ; from whence arise not alone the throat and lung diseases, so prevalent, but the whole class of female complaints, equally troublesome, if not so fatal. It is evident also that tight lacing, or wearing the clothing tightly fastened around the body, acts in the same injurious manner; and all the more in proportion as a lady is well formed; the Venus de Medici being the standard.

Females therefore suffering from these diseases, should be very careful to wear their clothing loosely about them, and if necessary, their skirts attached to waists made for the purpose, rather than tightly tied around them.

Delay of the First Menses.

Is frequently attended with bleeding of the nose; flushed face; dizziness and palpitation of the heart.

REMEDIES,—*Belladonna* and *Pulsatilla* one or two doses a day of each; if ineffectual follow with *Sulphur.* In persons of delicate health, with general debility, and loss of appetite, every thing that tends to promote the general health and strength will be of service; as daily baths of Alcohol and water; horseback riding, and free exercise in the open air.

Suppression of Menses.

REMEDIES.—If occasioned by wet feet or taking cold, *Pulsatilla.* If attended with fever, headache, pain in the back and limbs, add *Aconite*, and alternate every hour. If the face is red and flushed, and throbbing of the temples not relieved by Aconite, substitute *Belladonna.* If the congestion is principally in the chest, *Bryonia*, always continuing the *Pulsatilla.* The action of the medicines may be assisted by hot foot baths and warm drinks. If the suppression is long continued, accompanied with slight fever, pains in the hips and limbs, and especially if with bloating of the bowels or limbs, give *Apis Mell. Sepia* and *Sulphur* are also beneficial in cases of long standing.

Menses too Frequent.

REMEDIES. — *China*, a single dose every morning; and a dose of *Calcarea* and *Sulphur* at evening, on alternate days.

Menses too Profuse.

REMEDIES.—If very profuse and exhausting, with nausea; blood bright, *Ipecac*. If there is great weakness, faintness and ringing in the ears, *China*; or if with cold extremities *Secale*. Should these fail, or if it amounts to real flooding, give *Tincture Hammamelis* three or four drops every fifteen or twenty minutes; and inject the same by means of the female syringe; using a teaspoonful to a sufficient quantity of water. Should this check it too suddenly, causing flushed face and headache, a few doses of *Belladonna* and *Pulsatilla* will relieve.

Painful Menstruation.

REMEDIES.—*Pulsatilla* and *Cocculus*, either successively or in alternation every half hour till relieved. If with the pain there is nausea, coldness of the extremities, great debility, or diarrhœa, *Veratrum;* or if that is not sufficient, *Secale.*

If the difficulty is habitual, occurring at every monthly period, it can only be cured by appropriate treatment during the interval; for which a physician should be consulted.

Critical Period. Change of Life.

For the congestion to the head, flushed face, giddiness, headache and general debility common at this period, *Belladonna* and *Pulsatilla* are the most efficient remedies.

Leucorrhœa. Whites.

REMEDIES. — *Pulsatilla*, if the discharge is yellowish and thick; *Sepia*, if

4

acrid and excoriating. In obstinate cases use perseveringly *Calcarea* and *Sulphur*. Injections of Alum or Borax will afford relief.

DISEASES OF PREGNANCY.

Nausea and Vomiting.

Tabacum will generally relieve the most obstinate cases. If bilious symptoms predominate, the matter vomited bitter and green, with coated tongue, *Ipecac.* and *Nux vom.*

Heartburn. Acidity.

REMEDIES. — *Sulphuric Acid* in most cases ; follow if necessary, with *Nux Vom.*, *Pulsatilla* and *Sulphur*, giving Puls. or Sulph. in the morning and Nux at night.

Sleeplessness.

For inability to sleep at this period, use successively *Coffea*, *Belladonna* and *Nux vomica*.

Spots on the Face.

For the brown and yellowish spots sometimes appearing on the face during pregnancy, *Sepia* may be administered, followed if necessary, by *Sulphur*.

Constipation.

For this difficulty take *Sulphur*, a dose in the morning, and *Nux vomica* at night, and favor by a fruit and vegetable diet.

Varicose Veins.

REMEDIES.—*Pulsatilla* or *Lycopodium*, and bathe the swollen veins with *Tinc. Hammamelis*. If the veins become much enlarged, it is necessary to bandage the limb tightly.

Painful Urination.

REMEDIES.—*Cantharis*, if with painful burning and scalding; scanty and frequent.   Follow, if necessary, with *Pulsatilla* and *Nux vomica*.

Incontinence of Urine.

Frequently a mechanical difficulty caused by presence of the enlarged uterus upon the bladder ; *China* and *Nux vomica* may prove beneficial.

Preparation for Labor.

Can anything be given to shorten the period and alleviate the sufferings of childbirth ?   Dr. Hill, Prof. of Surgery at Cleveland, says in reference to this subject, " whatever others may think or say in relation to any preparatory treatment for labor I have reason to know as well as anything in medicine can be known, that patients thus treated, pass through labor much quicker, frequently in one-fourth the usual time. Their sufferings are much less, and the length of time for recovery to ordinary health after labor, is greatly abridged."

The treatment to which he refers is the use of two remedies, *Caulophyllin* and *Macrotin*, prepared from roots, and perfectly harmless in all cases; to be taken for two or three weeks previous to labor.— Having made repeated trial of them, the author would feel unwilling to dispense with them in his own practice.

Sore Nipples.

Where a tendency to this difficulty is known to exist, wash the nipples for a few weeks before confinement with Spirits of Wine or Brandy. If, notwithstanding, they become sore, apply a mixture of equal parts of Glycerine and Tannin, and take *Calcarea* and *Sulphur*.

Gathered Breasts.

When the breasts first become hard and swollen and the secretion of milk decreasing, give *Bryonia*. If there is much inflammation, fever and redness radiating from the centre, *Belladonna.*— If suppuration threatens, simmer together three or four ounces of Castor Oil and a handful of bruised Raisins; strain and apply on a piece of flannel. If suppuration cannot be avoided, give *Hepar Sulphur* and *Phosphorus*, and after the discharge, *Silicea* with poultices of Slippery Elm or ground Flax Seed.

Nursing Sore Mouth.

REMEDIES. — *Sulphuric Acid, Mercurius* and *Sulphur* will relieve most cases. Use the remedies successively, one at a time; a dose every three or four hours.

DISEASES OF CHILDREN.

Soreness of Skin. See "Chafing."

Jaundice.

Soon after birth, children sometimes exhibit a jaundiced appearance. It generally passes off in a few days without treatment. Should anything be required, a few doses of *Chamomilla* or *Nux Vomica* will remove the difficulty; if obstinate, *China*.

Snuffles.

Will be removed by a dose of *Nux vom.* at evening, followed by *Chamomilla* or *Dulcamara*, if there should be discharge from the nostrils with it.

Crying.

If the limbs are drawn up to the body, with flatulence of the bowels, or greenish

slimy stools, colic is the occasion.   Give
Chamomilla.   If the bowels are distend-
ed without evacuation, give *Colocynth.*
Should there be indications of earache,
raising the hand to the head, or redness
of the ear, give *Aconite* and *Pulsatilla.*
If starting suddenly from sleep, *Belladonna.*

Sleeplessness.

Often caused by irregularities of diet
and mental excitement of the mother.—
If not caused by Coffee taken by the
mother, give *Coffea.* If the child is restless
and feverish, starting suddenly from
sleep, *Belladonna.* If with fever and
dry heat, *Aconite.*

Inflammation of Eyelids.

Sometimes occurring soon after birth,
requires *Pulsatilla.*   If anything further,
Calcarea or *Sulphur.*

Colic. See "Crying."

Vomiting of Milk.

REMEDIES. — When excessive give *Ipecac, Nux Vomica* and *Pulsatilla;* or if the vomiting is sour, *Chamomilla.*

Sore Mouth. Thrush.

REMEDIES.—*Mercurius*, followed by *Sulphur* if necessary, and wash the mouth with a solution of Borax, or a little Borax and Honey.

Cholera Infantum.

REMEDIES.—If vomiting is a prominent symptom, *Ipecac.* If with great thirst, vomiting immediately after drinking, accompanied with profuse watery diarrhœa, *Arsenicum;* or if with sunken eyes, cold extremities, alternate with *Veratrum.*

This disease frequently runs a very rapid course, and unless speedily checked, a physician should be obtained without delay.

For the ordinary forms of Diarrhœa or Constipation, see these diseases elsewhere.

Difficult Dentition.

REMEDIES.—When teeth come very slowly; are attended with wasting of the flesh, and loss of appetite, *Calcarea* and *Belladonna*. For the diarrhœa, *Chamomilla, Ipecac.* or *Arsenicum* are most frequently indicated. " See Diarrhœa." If the bowels are constipated, *Nux Vomica, Bryonia* or *Opium.* For the fever, restlessness, and heat of head sometimes accompanying teething, *Aconite* and *Belladonna;* or if very wakeful and restless at night, *Coffea.*

Convulsions.

REMEDIES. — Give *Belladonna* and *Chamomilla* alternately every ten or fifteen minutes, five or six globules at a time laid upon the tongue, even though the child cannot swallow. Or *Ignatia* for spasms with twitching of the corners of the mouth, trembling of tho chin, and jerking of the body. *Cina*, where it is known to be caused by worms.

Immerse to the chest in a warm bath, keep the head cool, and if consciousness does not soon return, secure a free action of the bowels by injections of warm water with a little castile soap added.— As the disease may arise from a variety of causes, as the irritation of teething, worms, brain disease, suppression of eruptions, or disease of the stomach and bowels, the care of a physician is indispensablo to prevent the recurrence of the difficulty.

Water on the Brain.

The approaches of this disease are so insidious, that parents frequently do not recognise the danger, until too late to be remedied. Among the earlier symptoms are unusual peevishness, disposition to remain in a horizontal position ; rolling of the head from side to side, or boring in the pillow, with the head thrown back ; hot head with cold extremities ; sleeping with the eyes half open. Later, when nausea and vomiting set in, with blood-shot eyes or squinting, a rapid pulse, and a peculiar moaning cry, the disease is unmistakable. It is always a very dangerous and fatal disease, requiring prompt attention, and should never be left to family treatment. *Belladonna, Bryonia, Apis Mell.* and *Hellebore* are among the most reliable homœopathic remedies.

DIETETIC PREPARATIONS.

To assist invalids in a choice of proper nourishment, we add a list of various articles, with the best method of preparation. Simply remarking, that not all herein enumerated are proper for every form of disease; but as provision is made for every condition, from the most serious sickness through convalescence, as well as for the dyspeptic otherwise in usual health, selection should be made by advice of the attending physician.

LIQUID ALIMENTS.

Farina Gruel.

Mix a table-spoonful of farina in a little water; pour gradually on the mixture a pint of boiling water, stirring thoroughly, and boil for about ten minutes.

Indian-Meal Gruel.

Sometimes called water gruel. Sift the meal, and add three tablespoonsful to a quart of water; wash once or twice, changing the water as the meal settles; then boil for twenty minutes, stirring it constantly; strain and sweeten; or a little salt may be added.

Oatmeal Gruel.

Mix well two tablespoonsful of oatmeal with six of cold water in a basin; add this gradually to a quart of boiling water, constantly stirring it until sufficiently boiled, which will be in about ten minutes. Strain it and add a little salt. It may be pleasantly flavored by previously boiling a handful of raisins in the water to which the meal is added.

Sago Gruel.

Macerate a tablespoonful of sago in a pint of water, letting it stand in a warm place by the stove for an hour or two; then boil for fifteen minutes, constantly stirring it while boiling; may be sweetened, or flavored with a little lemon.—Sago is very nutritous and easy of digestion, and is well adapted to febrile and inflammatory complaints.

Sago Milk.

Is prepared by soaking a tablespoonful of sago in a pint of cold water for an hour; pour off the water and add a pint and a half of milk; boil slowly until the sago is well incorporated with the milk. May be sweetened, or where a stimulus is desirable, a spoonful of white wine added.

Tapioca Gruel, and Tapioca Milk.

Are made in the same way as sago gruel and sago milk, only tapioca being more soluble than sago, requires but half the time for its maceration and boiling.

Arrow Root Gruel.

Mix two tablespoonsful of Arrow root with water to a smooth paste. Gradually stir it into a pint of boiling water and let it cook till quite clear ; sweeten with loaf sugar. Milk may be used when preferable instead of water, especially for children.

Cracker Panada.

Pour a pint of hot water upon three or four crackers in a bowl; cover with a plate to confine the steam ; after standing till sufficiently cool, sweeten with white sugar, and flavor with a few raisins. The raisins are not to be eaten.

Bread Panada.

Place some very thin slices of stale bread, without the crust, in a sauce pan, and add rather more water than will cover them. Boil until the bread becomes pulpy; then strain off the superfluous water, and beat up the bread until it becomes of the consistence of gruel; and sweeten to the taste.

Crust Coffee.

Take a slice of bread a day old and toast without burning. Then put it in the oven and slowly roast it for an hour. Pulverize it, and pour upon it a pint of hot water; then add a little milk and sugar. Is an excellent substitute for coffee.

Barley Water.

Take two and a half tablespoonsful of pearl barley; wash it carefully with water, then add half a pint of water and boil for a few minutes. Throw away this water, and add four pints of boiling water; boil down to one-half and strain; sweeten and flavor with lemon, or where the bowels are sluggish two or three figs may be sliced and boiled with the barley. This preparation is nutritious and very digestible, and forms an excellent beverage in diseases of the bowels or urinary organs.   .

Rice Water.

Take of rice half a teacupful; wash it well; add two quarts of water, and let it boil for an hour and a half. Pour off the water and sweeten, or add a little salt.

Strong Beef Tea.

Take half a pound of lean, juicy beef; cut it in small pieces; put it in a bottle; add a pint of water, and cork the bottle loosely; set this in a kettle of cool water, and let it boil thoroughly till the meat is white and tasteless. Season with salt only, and add toast bread. If too strong add more hot water.

Beef Tea, No. 2.

Cut half a pound of lean beef in small pieces; add a pint of water, and set it over a slow fire; skim it as it boils.— When the beef is tender, strain off the tea, and season with a little salt.

Chicken Broth.

Wash half the breast and one wing of a tender chicken, put it in a sauce pan with a pint and a half of water, a little salt, and a tablespoonful of rice or pearl barley; let it simmer slowly, and skim it. When the chicken is thoroughly done, remove it from the broth. Chicken broth is of all others, least disposed to disturb an irritable stomach, and is therefore to be recommended in diseases of the stomach and bowels.

Mutton Broth.

Take a thick end of a loin of mutton; add a quart of water cold, to a pound of meat; add also a little rice or barley; let it boil slowly for three hours, carefully skimming off all the fat; season with salt only. If required in haste, take a piece of the neck or loin of mutton; cut it in pieces, and add a sufficient quantity of water; boil it quickly for an hour; skim it and season with salt as before; adding a little sago, rice, or barley.

Oyster Broth.

Open a half dozen large fresh oysters; put them with their liquor in a stew pan; place them over a moderate fire, and let them simmer slowly until they swell; remove them and strain off the liquor; if too fresh add salt as required.

Vegetable Soup.

Take one potato, one turnip, one onion, with a little celery; slice and boil in one quart of water for an hour; season with salt, and add to it toast bread. May be used where animal food is not allowed.

SOLID FOOD.

Farina.

Take three pints of milk; add slowly four large tablespoonsful of farina; boil from half an hour to an hour; put it in a jelly mould; and set it in ice or cold water to stiffen; may be eaten with cream sauce or sugar.

Farina prepared as gruel, or as above, is the most generally useful of all food in every kind of sickness; is nutritious, and readily digested.

Corn Starch.

May be prepared the same as Farina.

Rice Boiled.

Wash the rice; put it in a pudding bag allowing plenty of room for it to swell; put it in a kettle of boiling water, seasoned with a little salt. It will require boiling about two hours.

Rice Cups.

Take one quart of milk; three tablespoonsful of rice flour; two ounces of butter; put on the milk to boil; mix the rice flour very smooth with some cold milk; as soon as the former begins to boil, stir in the rice, and let it boil for twenty minutes. While the milk is warm add a little butter and salt. Rinse custard cups with cold water, and half fill them with the mixture. When cold they turn out of the cups and retain their form.

Rice Cakes.

Take two cups of rice, wash and boil over a slow fire, in three pints of water, until perfectly soft and clear. When done, mash it very fine; and season with a little salt. When cool, beat two eggs till light; add them to three half pints ot milk; then beat in by degrees six teacups of flour; add the rice, and after beating all well together, stir in a little saleratus; bake them the size of a breakfast plate, on a griddle, as buckwheat cakes.

Arrow Root.

Take of arrow root a tablespoonful; sweet milk half a pint; boiling water half a pint; boil for a few minutes; sweeten with loaf sugar; an excellent preparation for children.

Boiled Flour.

Take of wheat flour one pint; tie it up in a linen cloth as tightly as possible, and after dipping it in cold water, dredge the outside with flour till a crust is formed around it, which will prevent the water from soaking into it while boiling; it is then boiled until it becomes a dry hard mass. Two or three spoonsful of this may be grated, and prepared the same as Arrow Root. One of the best possible preparations in diarrhœa.

Unbolted Flour.

Take one tablespoonful of unbolted wheat, or Graham flour as it is called, mix it with cold water about as thick as cream; then stir it into one pint of boiling water, and let it simmer until it becomes perfectly clear. Stir in a little salt, and after beating it well, remove it from the fire, and add four tablespoonsful of cream, and sweeten with white sugar.

Tapioca.

Take three tablespoonsful of tapioca; wash, and cover it with water, and let it soak three hours; put as much more water to it, and boil until it is clear;

sweeten with white sugar, and add a little cream.

Sago.

Pick and wash the Sago, and to prevent the earthy taste, soak it in cold water for an hour or two. Pour off the water; add more and boil gently until it becomes clear; sweeten and flavor as desired.

Cracked Wheat Mush.

As the wheat swells very much in boiling, it should be stirred gradually in boiling water, until a thin mush is formed. Then continue to boil moderately for one or two hours. If ground very coarse, it will require much more time to boil thoroughly.

Rye Meal Mush.

Prepared the same as cracked wheat. Is particularly adapted to those suffering from habitual constipation, as it is quite laxative in its nature. May be eaten with syrup or sugar.

Oatmeal Mush.

Prepared the same as above.

Brown Bread.

Graham Bread as commonly known; or Dyspepsia Bread. Is extremely use-ful in cases of habitual constipation. It may be made in the following manner: Separate the coarser particles of the flour by passing it through a common hair seive. Take six quarts of the flour, one tea cup of yeast, and half a tea cup of molasses; mix these with a pint of milk warm water, and a teaspoonful of saleratus; make a hole in the flour, and add the other ingredients; stirring it in the middle of the meal till it is like batter; when sufficiently fermented, make the dough into four loaves, which will weigh when baked about two pounds each. It requires a hotter oven than fine flour bread, and must bake about an hour and a half. If saleratus is considered objectionable, it may be made without it.

PUDDINGS FOR CONVALESCENTS.

Tapioca Pudding.

Beat the yolks of two eggs together, add half an ounce white sugar, and stir the mixture into Tapioca Milk, (which see.)

Arrow Root Pudding.

Made the same as Tapioca.

Boiled Bread Pudding.

Grate about half a pound stale bread; pour over it a pint of hot milk, and leave it to soak for an hour in a covered basin;

then beat up two eggs and mix well. Put the whole in a covered basin just large enough to hold it; tie it in a cloth and place it in boiling water for half an hour. May be eaten with sugar sauce.

Rice Pudding.

To five tea cups of milk add half a tea cup of rice, and one tea cup of sugar; bake for three hours over a slow fire.

JELLIES.

Calves Foot Jelly.

Take two Calves' feet; remove the bones; divide them into pieces, and throw them into warm water to soak out all the blood; after this wash them well in cold water; then add one gallon of water and boil down to a quart; strain it and when cold remove any fat remaining; add to this the white of six or eight eggs well beaten; half a pound of loaf sugar and the juice of four lemons, well mixed. Boil the whole for a few minutes, stirring constantly, and then strain through flannel. If wine is added let it be put in with the sugar and eggs.

Currant Jelly.

Mash the currants and strain; to every pint of juice add a pound of loaf sugar; boil it till it jellies. Skim it while boiling, and put in cups while warm. If desired to retain its form, add Isinglass dissolved in warm water to the juice before boiling, in the proportion of half an ounce to a quart of juice.

Isinglass Jelly.

Take of Isinglass two ounces; water two pints; boil it to one; strain, and add milk one pint; white sugar candy one ounce. This is one of the best articles of nourishment for children in cases of Cholera Infantum.

Rice Jelly.

Boil half a teacupful of the rice, and a teacupful of white sugar in a pint of water, until it becomes thick and jelly-like. Flavor with a little lemon or orange water. Invaluable in summer complaints.

Tapioca Jelly.

Wash the tapioca thoroughly, allowing it to steep four or five hours, changing the water two or three times. Take of it two tablespoonsful; water one pint; boil gently for an hour, or until it assumes a jelly-like appearance; sweeten, and flavor with lemon juice or a little wine.

Wine Jelly.

Boil and clarify half a pound of loaf sugar; dissolve one ounce of Isinglass in a very small quantity of warm water, and strain it into the syrup; when nearly cold, add half a pint of wine; mix it well, and pour into a mould or bowl. Is very nutritious for convalescents.

Irish Moss Jelly.

Wash the Moss two or three times in cold water, to remove the salt taste; add half an ounce of it to a pint and a half of fresh milk; boil down to a pint; strain and sweeten as agreable; flavor with the juice of a lemon; or a little wine. It may be made also by using boiling water instead of milk; simmering it until the mass becomes thick and pulpy.

Biscuit Jelly.

Take of white biscuit four ounces; water four pints; boil down one half; strain and evaporate to one pint; add white sugar one pound; red wine a gill. Useful in debility of the digestive organs.

Arrow Root Blanc Mange.

Mix two tablespoonsful of Arrow Root with a little cold milk. to the consistence of cream; stir it into one quart of boiling milk; sweeten, and flavor with lemon; let it boil, and continue stirring until it is quite thick and smooth; pour it into a mould, and set aside to cool; may be eaten with cream and sugar; or when directed, flavored witn a little wine.

BEVERAGES.

Cold water is the most grateful and
desirable in almost all cases of sickness;
especially in fevers and inflammatory
diseases; should be used sparingly in
bowel complaints, and only at a natural
temperature; never with ice, though ice
may be broken up in small lumps, and
eaten as a substitute for water, particularly where the stomach is too irritable
to bear liquids. *Crust Coffee* made as
directed above is both palatable and
nourishing. Also *Rice Water*, made by
boiling rice with water, and pouring off
the liquid when sufficiently done. *Lem-
onade* when not contra indicated by looseness of the bowels, or the use of special
medicines, is one of the most cooling
and refreshing drinks, especially in
fevers. Other preparations are:

5

Tamarind Water.

Fill a tumbler one-third full of Tamarinds; fill up with cold water, cover it, and let it stand for half an hour. Very refreshing in fevers.

Prune Water.

Put a handful of good French Prunes in a bowl; nearly fill with warm water; cover, and let it stand till cool; if preferable sweeten with white sugar. One of the best possible preparations where the bowels are constipated.

Jelly Water.

Dissolve Currant or any other Jelly, in cold water.

Gum Arabic Water.

Pour a pint of hot water on an ounce of Gum Arabic. If allowed, add the juice of a lemon, and sweeten.

Flax Seed Tea.

A most useful and agreable drink; made by steeping an ounce of unground flax seed in a pint of boiling water; cover the vessel closely; and let it stand an hour; strain it; add the juice of a lemon, and sweeten. Promotes expectoration in bronchial catarrhs.

Slippery Elm Infusion.

May be prepared in a similar manner to flax seed.

Cocoa.

Take a teacupful of ground Cocoa; put it in a small bag made of very thin muslin, and tie it close; put it in a pot with three half pints of boiling water and one pint of boiling milk; boil the whole for half an hour; then pour it off into another vessel.

Milk.

Is useful when a nutritious but not stimulating diet is desired. It is especially so, in chronic inflammation of the chest, bowels, and bladder ; in consumption and in diseases of children ; with whom it always agrees better than with adults. With some it proves heavy and difficult of digestion owing to the butter contained in it ; with such skimmed milk will agree better ; boiled milk is preferable in bowel complaints.

Wine Whey.

Boil a pint of milk, and when boiling add a wine glassful of Madeira or other wine ; let it boil again, and then remove it from the fire, and let it stand a few minutes without stirring. Then remove the curd, pour the whey into a bowl and sweeten it. This is required in some prostrating diseases as Diphtheria, but should not be used except by advice of the physician.

Milk Punch.

As prescribed in incipient lung disease and after hæmorrhages, is best prepared by adding to a glass of milk, a fresh raw egg thoroughly beaten to a froth; mix well and sweeten with white sugar; then add from half to a full wine glass of good malt whiskey.

GENERAL INDEX.

List of Books Published by E. Darrow & Brother,

65 Main St., Rochester, N. Y.

Price.

Elements of English Grammar—by Wm. C. Kenyon, A. M., Pres. of Alfred University—8th ed., revised,	.60
Morey's Practical Arithmetic,	.38
Scientific Agriculture; or the Elements of Chemistry, Geology, Botany and Meteorology. applied to Practical Agriculture, with 45 Engravings—by M. M. Rodgers, M. D. 2d ed. enlarged; muslin,	.75
Physical Education & Medical Management of Children—by M. M. Rodgers, M. D., with engravings,	.25
Differences between Old and New School Presbyterians—by Rev. Lewis Cheeseman, D. D.,	.75
The Bridal Keepsake—by Mrs. Coleman; a neat miniature, steel eng., gilt edge,	.38
Darrow's Workman's Pocket Time Book—Cap. 8vo., various styles,	
Sab. School Sec'y's Register, & Librarian's do.; each	.50
" Teacher's Class Book,	.12
Fireside Gift, or Lectures for the Fireside, founded on the Ten Commandments—by Dr. W. A. Alcott,	.75
Arithmetical Chart, in Two Numbers—by J. Homer French; for the use of Schools & Academies, each 36 by 54 inches in size; neatly mounted and varnished, with cloth backs. For the set.	4.00
Woodworth's Cabinet of Curious Things: embellished with 100 engravings—by Francis C. Woodworth,	1.25
Tupper's Proverbial Philosophy & Thousand Lines,	.75
Elwood's Grain Tables: giving the Price of all kinds of Grain, in dollars and cents,	1.25
Christian Gift—by Rev. F. De W. Ward,	.25
Manual for the Medical Practitioner—by Gerhardus Arink, M. D.; 8vo., muslin,	1.00
Church Music: Presbyterian Hymn and Tune Book,	.50
Daily Walk with God—an Essay by Rev. S. Porter,	.25
Summer Vacation Abroad—by Rev. F. De W. Ward,	.75
" " " Gilt	1 00
Erna, the Forest Princess—by Gustav Nieritz; 18mo.	.33
Slavery Unmasked—by Rev. Philo Tower,	1.00
"Ministerial Legacy,"—by Mrs. Powell; 3 steel engr.	1.00
Martin's Interest and Average Tables, large 4to.,	2.00
Doyle's Ready Reckoner,	.25
Darrow's Diaries; all styles—yearly	

We Manufacture Cornell's Improved Globes.

www.ingramcontent.com/pod-product-compliance
Lightning Source LLC
Chambersburg PA
CBHW021932190326
41519CB00009B/1000